The Day the Sun Went Out

by Pia Lord

The Day the Sun Went Out

by

Pia Lord

ISBN-10: 1539813037
ISBN-13: 9781539813033

DEDICATION

To J.C. who teaches me that patience and faithfulness are two of the greatest virtues in life. I hope this book can interest children and adults to enjoy solar science and seek to know more about the universe.

CONTENTS

ACKNOWLEDGMENTS

The scientific endeavour to understand happenings in our universe, ones seen and unseen, bring mystery and wonder to our explorations. The solar eclipse is one of the many wonders of the sun in our solar system. I'd like to thank the helio-physicists that each day bring us new information about stars, especially our own. I appreciate the freedom of the internet in knowledge sharing which adds to our enriched society. Albert Bierstadt's landscape painting has been an inspiration for its beauty and magnificence.

The Day the Sun Went Out

by Pia Lord

The Alpine Mountain Lake

One sunny Saturday, Alexander and his parents had gone for a hike in the mountains. They had been walking for about three hours, enjoying the beautiful landscape of streams, trees and alpine region flowers, when they came upon a lake. The water was crystal clear, reflecting the blue green of the surroundings. It reflected the entire sky just as a mirror would do on a warm sunny day.

Just then the day went dark! It was like someone had turned off the light! As if he sun went out and they could barely see their footsteps in front of them. Alexander ran over and said, " Mommy, Mommy why is it dark out? Is it night time so soon?" "No, Alexander, it is just now coming on the middle of the day. It is about noon. Something has happened to the sun. What do you think it could be?

What Happened to the Sun?

How can we find out? Perhaps we will ask Daddy what he thinks?" she replied.

But Alexander's father had walked on ahead to see what the path was like saying that he would rejoin them in about a half of an hour.

"Daddy is up ahead!" said Alexander reminding his mother.

"Well we have other resources don't we! We can also inquire of the computer search engine to find out more about this," said his mother.

She then took out her internet linked mobile phone with hotspot and they sat down on the grass. Into the search box she typed "sun going out" for keywords. After scanning over a few of the returns, they clicked on one. They learned that some scientists think that the sun is in a cooling period and shrinking by about 10km per year. Other astronomers or helio-physicists say just the opposite, meaning that the sun is expanding!

"This still does not answer our question, however, as to why the sun is eclipsed right now!" exclaimed Alexander to his Mommy.

The New Moon Eclipses the Sun

"Let's search some more", agreed his mommy. "The word "solar" means sun so let's use keyword "solar". Do you remember the word "eclipse" from our story on the lunar eclipse? Let's search eclipse too and see what we get."

So they put "solar eclipse" into the search engine and clicked enter. Up came lots of information about solar eclipses. With their flashlight app, they quietly read about the solar eclipse. Then they looked up to the sky again to see what else had happened.

As they were waiting, sitting and searching, the sun had gone through some changes. They noticed the sun now had a bright ring around it but its center was dark. It nearly looked like a diamond ring! It had a big bright shining light emerging from one side, appearing just like the stone of a large diamond ring.

While the minutes passed though, it was still noticeably changing!

The Diamond Ring Effect

The Sun was being hidden by the moon as it passed in front of it. It was the most interesting thing! Alexander was just thrilled to see such a cool sight on his hike in the mountains. He wondered if his Dad was seeing it too. He hoped that his Dad would be back soon from walking on ahead. But it was too dark for Alexander to go run and find him on the path, away from his mom.

"Look Mommy," cried Alexander, "the sun is coming back again. It is getting brighter and the moon is right next to it!"

The Moon Appeared Next to the Sun.

Fortunately, in the seven minutes that they were resting and searching the internet, the sun had returned to its good-old, warm, shiny self. Alexander was so relieved.

It was just that day Alexander began to take more notice of the changing sun. Little Alexander pondered this. Why had it gone out so suddenly? The big round light in the sky had darkened, he wondered why. He then realized that the sun going out could have very serious consequences for life, his life and all life on Earth. He knew he had to do more to understand what was going on.

The next day his cousin Max came over.

Max said, " Did you notice there was no sun for part of the day yesterday?"

Alexander said, " Yeah, it just went out! How can it do that? We need the sun for lots of things. We need to know why the sun blinked yesterday!"

"This could be very serious for life on Earth," agreed Max.

"I know. We need to do something about this. We don't want many days like this," said Alexander.

"But what, what can we do?" asked Max.

"My Mom and I looked up "solar eclipse" on the computer yesterday. We found out that the new moon is going right in front of the sun and blocking it out. But I still want to know more about it," said Alexander.

"I want to know more" said Alexander.

"My Dad has a friend who is an astronomer. He works at Kepler Labs. That is a part of NASA, the National Air and Space Administration. That's where they study all things about space. He told me they have a big telescope there. Maybe we can find out more about it. Let's call Kepler Astronomy Labs and talk to an astronomer, like my Dad's friend," said Alexander.

"Great idea! Maybe they can help," said Max.

Max dialed the number on his cell phone.

"Hello this is Max. May I please speak with an astronomer who knows about the sun and solar eclipses?"

"One moment please" said the operator.

"Hello, this is Professor Tycho. How may I help you?" asked the Kepler astronomer.

"Hello Professor Tycho. We have a big problem that we hope you can help us solve. Yesterday, the day went dark when the sun went out. We are afraid the whole world will be dark forever at random times, if we don't figure out what happened. We know that the sun has powerful solar storms which cause solar weather on Earth and surely elsewhere in our solar system.

Life here on Earth depends upon a constant sun to provide light and warmth for life. We need it to grow all of our fruits and vegetables. Other forms of life and our health also depend upon it. We want to make sure that it is alright and can shine always brightly. Can you send one of your spaceships to the sun to help us understand why the sun went out all of a sudden? Please?" asked Max very politely.

Then Professor Tycho chuckled in a friendly way and said, "Boys, boys we will be delighted to help you. We have international space station photos of exactly what you saw. The astronauts take pictures from up in space. These pictures are taken from a different angle than what you see here on Earth".

"On Earth we see the sun being blocked by the moon in a solar eclipse. The sun passes behind the moon and it looks like a diamond ring, since only one edge of the sun is still visible. In fact it is called the "diamond ring effect". Then the sun continues completely behind the moon. At this moment the sky totally darkens and we have the total solar eclipse. When the sun emerges again, the "diamond ring" effect happens one more time. This can take up to several minutes or longer," said

Professor Tycho.

"NASA also sent a team of scientists to the Libyan dessert to study the solar eclipse that happened in 2008. These scientists were making observations of the total eclipse for their own solar research studies. The dessert was the best place on Earth to view that particular solar eclipse," added Professor Tycho.

August 21, 2017 there will also be another solar eclipse visible from the northern hemisphere of the North American continent. Here at NASA we are also flying a number of satellites which point directly to the sun. Heliosat is our newest solar observing satellite. I will contact the control headquarters and find out any information they have on this recent solar event. You can also check the website. NASA Solar System for Kids has fun games and lots of great information.

The Solar Eclipse with the Diamond Ring Effect

However, boys, I want to tell you one little secret. How old are you?" asked Professor Tycho.

Max said, "Ten years old" and Alexander proudly said, "I'm six."

Professor Tycho then said, "There are some amazing things going on around our universe. So as you know now, as the Earth and Moon revolve, or go around the Sun, a solar eclipse can sometimes happen. The total solar eclipse is when the moon passes in front of the sun, in the new moon phase when it is round and full. As such, it can completely block out the sun. it can be dark for up to about 7 minutes or so, "said Professor Tycho.

"That's what my Mom, Dad and I saw yesterday out in the mountains when we were going for a family hike!" exclaimed Alexander.

"Does it happen often?" asked the boys then chiming in together.

"In the northern hemisphere in North America, to see a total solar eclipse happen is in fact quite rare. Numerous eclipses have been seen from the southern hemisphere such as the one seen in 2008 in the Libyan dessert. You were very fortunate to see the eclipse. When the moon moves further in its orbit, the sun gets uncovered again an light from the sun once again is visible. So really what you are seeing is not the sun going out but one of a number of different types of solar eclipses. There are also partial eclipses. The word eclipse means to cover or block out. If something is eclipsed it is put into a shadow. In this case it is the sun that is eclipsed by the moon," said Professor Tycho.

"So you mean the Earth won't stay dark forever because the sun did not disappear," said Alexander very relieved.

"Right Alexander! A partial solar eclipse may happen a few times a year but then only lasts a short time," added Professor Tycho. "When the annular eclipse has the ring effect around it what you are seeing is the solar corona, the outer region of the sun. Corona comes from the word crown. So it's like a crown that the sun is wearing."

"Well boys, I need to get back to work now. Keep up that curiosity about the universe! There is so much more to know and find out!" said Professor Tycho as he hung up the phone.

That day after lunch, Alexander's mom said, "Alexander and Max, it is a great day to play outside. So run along now. Remember never to look at the sun through a regular telescope or binoculars. You will damage the retina of your eye.

The sun won't be eclipsed today. Remember also to wear your protective glasses while in the sunlight and never look directly into the sun."

"Okay its now off to play in the back yard for you two!" said Alexander's Dad.

They looked at each other, did a high five and said, "We did it, we did it, we did it!"

They ran outside and found the sun shining brightly.

"Alexander said, "I am so happy that it is a bright, warm and sunny day!"

Max agreed, " I like to know that the raspberries on the bush will continue to grow in the sun's warmth and light. Let's go get some."

So they ran over to the raspberry patch in Alexanders backyard that had been growing there for ten years. They picked handfuls of raspberries and popped them into their mouths. Afterwards, they ran to the backyard and climbed all over the gym, swung from the tree ropes and climbed up the climbing wall. They pretended to be Tycho Brahe and Johannes Kepler sighting new suns, planets and moons through imaginary telescopes!

Alexander and Max, astronomers of the future

The

End

GLOSSARY TERMS

Annular Solar Eclipse-An eclipse in which the moon covers only a central portion of the sun, leaving the corona to shine like a ring. Solar eclipse in which the shadow that the new moon casts cannot create a total eclipse due to the distance of the moon from the Earth in its orbit around Earth. The moon in an annular eclipse is closer to its apogee.

Apogee-The location of a planet or moon in its orbit to its primary body when it is at its furthest point.

Tycho Brahe- Danish astronomer, Royal Mathematician to the Danish Court, who kept meticulous records based upon keen observation of the night sky.

Corona- the gaseous outer regions of the sun only visible during a period of a total eclipse.

Eclipse- to block out or to cast into shadow

Johannes Kepler- found three Laws of planetary Motion based upon the meticulous work of Tycho Brahe. Was able to determine the position of planets in the future, predict the transits of Venus and Mercury using mathematical and logarithmic formulas that he devised.

Kepler Labs-The first space mission which searches for Earth-sized as well as smaller planets in the habitable zone of other stars in our neighborhood of the galaxy. kepler.nasa.gov

Partial Solar Eclipse- A solar eclipse in which the moons shadow is seen to cover only part of the disk of the sun due to the viewer location on Earth.

Penumbra- A partial shadow, as in an eclipse, between regions of complete shadow and complete illumination.

Perigee- The location of a planet or moon in its orbit to its primary body when it is at its closest point.

Photosphere-the lowest layer of the sun visible from Earth.

SOHO-Solar and Heliospheric Observatory is a satellite which orbits the sun, studying it on a 24 hour basis. It is the main source of our information on solar storms and solar weather. https://sohowww.nascom.nasa.gov/home.html

Solar- of or relating to the sun, or meaning sun.

Total Solar Eclipse-A perfect alignment of the Earth, New Moon and Sun in which the shadow of the new moon cast on the Earth is sufficient to cover the entire disk of the sun. The total solar eclipse occurs when the moon is closest to perigee in its orbit about the Earth.

Umbra- the darkest shadow cast by the moon in a solar eclipse.

ACTIVITIES

1. Using a cell phone camera hold the video up to the sun. Run 30 seconds of video capturing the rays of the shining sun. Discuss what you see on the video. Answer the following questions.

    a. Do the rays of the sun represent particles or waves?
    b. Is it possible to distinguish individual photons?
    c. What do you see on the video?
    d. Do you think that individual photons are present on the digital video?
    e. Draw a picture of the particles or waves that you see on the video when you press pause.

2. Obtain a telescope with a solar filter. Only perform this activity if you have a properly fitted solar filter otherwise you can do significant damage to the retina of your eyes. Once set up, locate the sun through the solar filter. Answer the following questions.

    a. On a sheet of paper label sun layers, describe what you see.
    b. Can you see any spots on the ball of the sun?
    c. Add these to your drawing in the correct place.
    d. Label these sunspots
    e. Repeat this activity every week and study changes in your drawings over time. Are there changes? Why or why not

RESEARCH PROJECTS

1. Students can choose one topics related to the sun from this group: solar corona, sun spots, solar eclipse, fusion vs. fission, magnetism on the sun, 11 or 22 year solar cycles, solar telescopes, solar weather, solar flares, human health, communications and the sun or another solar related topic of their choosing. Students research to write a 5-10 page paper, listing 10 sources, with 2-5 illustrations, and in depth explanations of each solar phenomena.

2. Research different solar satellites. Some examples are listed in the glossary of this book such as SOHO. Find other examples. Select one and write up a technical description of the parts and how they function. Draw a diagram and label the parts of the satellite. Give a thorough description to increase your understanding of how a satellite is built, its parts and how it functions. Students research to write a 4-6 page paper, with at least 3 illustrations and functional explanations of the parts.

REFERENCES

The Day the Sun Went Out by Pia Lord Publish America 2012

https://sohowww.nascom.nasa.gov/home.html

http://www-ssg.sr.unh.edu/index.html?tof/Missions/Ace/Model/acemodel.html

http://www.mreclipse.com/Special/SEprimer.html

http://www-ssg.sr.unh.edu/index.html?tof/Missions/Ace/Model/acemodel.html

http://seal.nascom.nasa.gov/explore/

https://sohowww.nascom.nasa.gov/classroom/elem_poster09_allweb.pdf

https://sohowww.nascom.nasa.gov/classroom/

https://sohowww.nascom.nasa.gov/classroom/for_students.html for further exploration.

ABOUT THE AUTHOR

Pia Lord has written 19 books including short stories, illustrated children's science stories, supplemental science classroom books, poetry, haiku, open water swimming guides, and science fiction novellas.

The Pia Lord Company began in 2006 with the publication of her first book Harvest While the Orchard is Aplenty by Olympia Fiedler. Her books can be found at **www.pialord.com** as well as at the Amazon **Pia Lord Author Page**.

Pia enjoys traveling, swimming, skiing, cycling and music. She is married, has one son and lives in NJ, USA.

Science Books By Pia Lord

Cato the Caterpillar
The Night the Moon Went Out
Let's Take a Trip in Our Spaceship
El Teide-Canary Island Volcano
The Masca Gorge
Tenerife, Canary Islands, Spain
The Upper Limit